ISBN 978-1-334-07323-6
PIBN 10551426

1 MONTH OF
FREE
READING

at
www.ForgottenBooks.com

By purchasing this book you are eligible for one month membership to ForgottenBooks.com, giving you unlimited access to our entire collection of over 700,000 titles via our web site and mobile apps.

To claim your free month visit:
www.forgottenbooks.com/free551426

THE AMERICAN OR EASTERN OYSTER

Cover Picture: Oystermen picking out legal sized oysters from material collected from public beds. Small oysters and old shells and their fragments are returned to the beds.

UNITED STATES DEPARTMENT OF THE INTERIOR
Stewart L. Udall, *Secretary*
Frank P. Briggs, *Assistant Secretary for Fish and Wildlife*
FISH AND WILDLIFE SERVICE, Clarence F. Pautzke, *Commissioner*
BUREAU OF COMMERCIAL FISHERIES, Donald L. McKernan, *Director*

The American or Eastern Oyster

By

VICTOR L. LOOSANOFF

Circular 205

Washington, D.C.
March 1965

CONTENTS

Page

Introduction.. 1
Environment.. 1
Anatomy and physiology... 2
Growth 8
Reproduction
    Gonad development and spawning... 8
    Eggs and larvae.. 12
    Effects of temperature on eggs and larvae................................ 14
    Effects of salinity on eggs and larvae................................... 15
    Effects of turbidity on eggs and larvae.................................. 16
    Food and feeding of larvae... 17
    Metamorphosis, or setting, of larvae..................................... 17
    Diseases of larvae... 17
Oyster enemies
    Diseases and parasites... 17
    Predators 20
    Competitors 25
Oyster industry.. 28
Sanitary control 33
Selected references 34

The American or Eastern Oyster

By

VICTOR L. LOOSANOFF, Senior Scientist

Bureau of Commercial Fisheries
Tiburon, Calif.·

INTRODUCTION

The American or, as it is more often called, Eastern oyster is the oyster of commerce of our Atlantic and Gulf of Mexico coasts and is also sold in small quantities on the Pacific coast. Its scientific name is Crassostrea virginica, and it is a true oyster, being a member of the Phylum Mollusca, Class Pelecypoda, and Family Ostreidae. True oysters are distinguished by having dissimilar lower and upper shells, by attaching the left shell to a substratum, and by having no traces of foot and byssus in adults. Their shell ligament is a band between the two valves which may be of triangular shape.

Altogether, more than one hundred living species of oysters have been described, but only a few are of economic importance. Most oysters occur between tidal levels or in shallow waters of estuaries, but some species live in depths of several thousand feet. They are encountered along the temperate and tropical coasts of all continents.

In addition to the Eastern oyster, two other commercial species are cultivated in this country. One is the Japanese or, as it is now called, Pacific oyster, Crassostrea gigas, grown on the west coast, principally from imported seed; the other is the Olympia oyster, Ostrea lurida, a native of the Pacific coast. In 1949, the European flat oyster, O. edulis, was introduced into New England and now is found occasionally in Maine waters. Recently, small numbers of hatchery-grown oysters of this species have been used in planting experiments in California.

The American oyster (fig. 1) is widely distributed and in some areas is extremely abundant. It is found from Massachusetts south along the eastern coast of the United States and also along the Gulf of Mexico coast. Some groups still live in the waters of Maine and New Hampshire where they were considerably more abundant several decades ago. The American oyster is also cultivated in Canadian waters, principally in the shallow southern part of the Gulf of St. Lawrence.

The Pacific oyster is imported as seed from Japan and grown in large quantities in Puget Sound, Willapa Bay, and Grays Harbor in Washington, and also in Humboldt, Tomales, and Drakes Bays of California. Small quantities of Pacific oysters are also grown in certain protected inlets along the Oregon coast. It was first imported to the United States on a commercial scale in 1905, but during the last 3 decades its production has rapidly increased and now constitutes about 15 percent of the total annual yield of oyster meats in the United States. It is a large, rapidly growing mollusk, anatomically and in general appearance closely resembling the Eastern oyster.

The Olympia oyster, a native of the Pacific coast, is found from Charlottestown, British Columbia, to San Diego Harbor, Calif. It occurs in greatest numbers in Washington, especially in the lower part of Puget Sound.

ENVIRONMENT

The American oyster (hereafter referred to only as oyster) is adapted to live in waters with considerable variations in salinity and temperature. Its optimum salinity range is roughly from 10.0 to 28.0 parts per thousand (p.p.t.) or, in other words, in water containing about 1.0 to 2.8 percent sea salt. The oyster can survive in the open ocean for some time, but usually it does not reproduce or grow well there. It also can survive periods of spring floods or heavy rains when the salinity of the water is abnormally reduced. In such instances, however, the temperature is an extremely important factor in survival because the lower the temperature, the longer the oysters can live in water of low salinity. For example, experiments have shown that when the water temperature is only about 50° F. many oysters can survive exposure to a salinity of 3.0 p.p.t. for 30 days.

Oysters in Long Island Sound, where salinity of the water is about 28.0 p.p.t., fed even when placed in water of reduced salinity, sometimes

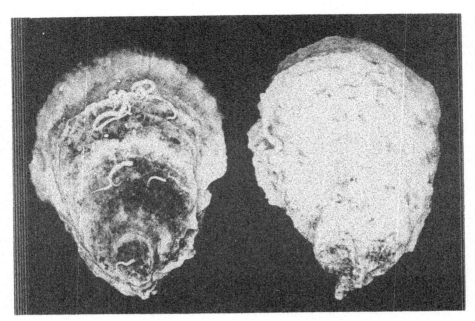

Figure 1.--American or Eastern oyster, *Crassostrea virginica*, photographed to show its upper and lower shells. New shell growth is clearly seen along the edges. The white tubes on the shells are some of the fouling organisms normally associated with oysters.

as low as 5.0 p.p.t., although under such conditions their behavior was often abnormal and they did not grow. However, in water of 7.5 p.p.t. the oysters grew, but much more slowly than at higher salinities. The lowest salinity at which spawn of Long Island Sound oysters will develop normally is about 7.5 to 10.0 p.p.t. However, some oysters with ripe gonads which developed at an optimum salinity of about 25.0 p.p.t. spawned when placed in a salinity of only 5.0 p.p.t.

Experiments also have shown that these oysters can withstand sudden changes from low to high salinity, and vice versa, without serious injury. They fed and expelled true feces within a few hours after the change. In general, oysters accustomed to low salinities stop feeding and close their shells at lower salinities than oysters conditioned to live in water of higher salinity. As a rule, changes in salinity of oyster body fluids closely follow changes in salinity of the surrounding water.

The temperature range that oysters can tolerate also varies greatly. For example, in Long Island Sound the winter water temperature over the oyster beds may drop below 32° F., while during the warm part of summer it rises to about 73° F. in deep water and as high as 78° F. in inshore areas. Oysters in certain shallow water areas of Chesapeake Bay withstand seasonal temperature variations from the freezing point in winter to 90° F. in summer. In the Gulf of Mexico the annual range of temperatures over the oyster beds is from about 50° to 90° F.

It is interesting that the oyster, living between tidal levels, may be frozen solid in winter; yet, if not disturbed, will thaw out and survive upon being covered by water. If a frozen oyster is shaken or dropped, however, changes occur in the cells of its body that lead to death of the mollusk.

ANATOMY AND PHYSIOLOGY

Various aspects of anatomy and physiology of oysters are described in hundreds of articles written by different authors. This knowledge, however, is very well summarized in the book, "Oysters," by C. M. Yonge, which is an important general reference. Most of the following material on anatomy and physiology of oysters is based on Yonge's summary.

The oyster shell consists of two valves held together at the hinge by a complex elastic ligament. The upper valve normally is flat, while the lower is concave, providing space for the body of the oyster. The concave shell is the one by which the oyster is normally attached. The valves fit closely together, making a watertight seal when the oyster closes, provided the edges of the shells have not been broken off or otherwise damaged.

The shell is principally limestone (calcium carbonate) and, therefore, is quite heavy. The formation and repair of the shell are functions of an organ called the mantle, which surrounds the body of the oyster (fig. 2). The umbo, at the hinge end, is the oldest part of the oyster shell. As the oyster grows, the mantle secretes successive layers of shell material, each projecting beyond the previous one. This secretion finally results in a succession of concentric lines marking the external surface of the oyster shell.

Basically, the oyster shell consists of three layers. The thin outer layer of organic nature, known as the periostracum, protects the calcareous shell during its formation. Because it is thin and sometimes quite eroded, this layer is often overlooked. The inner epithelium of the outermost fold of the mantle edge secretes the oyster's periostracum.

Under the periostracum are second and third layers of shell. Both have an organic matrix but are largely calcareous. The second is the prismatic layer, which consists chiefly of crystals of calcium carbonate, a limelike substance. The prismatic layer, secreted by the outer epithelium of the outer fold of the mantle edge, also shows definite concentric markings of successive periods of growth. The third or innermost layer is sometimes called the nacreous layer and consists principally of thin sheets of calcium carbonate covering the inner shell surface. This surface usually is smooth and white except for an area of purple scar where the shell muscle is attached. The pearly inner layer of the shell is laid down by the entire mantle surface and, as a result, is lustrous and smooth, showing no growth rings.

The ligament, which is formed by a specialized gland at the edge of the mantle, consists of the same three shell layers but with the periostracum always worn away and the other two layers not calcified. It is continually being eroded above and being added to below, the latter process being the greater so that it thickens with age.

The chemical composition of oyster shells may vary from area to area and sometimes with individuals of different ages. In general, however, the oyster shell contains from 93 to 95 percent calcium carbonate and about 0.5 percent organic matter. Magnesium carbonate and calcium sulfate also are present in small quantities. Old oyster shells are used for a variety of purposes, including chicken feed. Most of them, however, are saved and replanted later on the oyster beds to "catch" the new generation of oysters.

The main sections of the soft part of the oyster can be seen if one of the shells is removed (fig. 2). A thin, creamy, membrane-like organ called the mantle lies against the inner sides of both valves. The two principal functions of the mantle are to protect the more vital organs and to secrete the shell.

Pearls are produced by the oyster mantle, but in the American oyster they are valueless because they lack lustre and usually are misshapen. Pearls are small deposits formed around a nucleus, such as the cyst of a parasite, a particle of broken shell, or even a grain of sand. Usually pearls are embedded in the mantle or are found just under the surface of the meat; however, they are occasionally formed as blisters attached to the inside surfaces of the shell.

Sometimes thin layers of greenish or brownish color are found embedded in the calcium material of the inner shell surface. These layers, which differ in size, are of organic nature, being composed of material known as conchiolin. These layers may be secreted by the oysters as a defense against the intrusion of boring sponges, worms, or other enemies attacking the shells. These layers, however, may be found in oysters not infested with these forms.

A common characteristic of the oyster shell is the occurrence of soft "chalky" material embedded in the harder inner layers. These deposits may be used by the oysters merely as a measure of economy in smoothing out the inner surfaces of the shells, because depositing "chalky" layers requires only one-fifth as much material as is needed to build the same volume of shell of the usual hard, subnacreous layers. Normally, these deposits are laid down in summer and later in the season are covered with harder layers.

The mantles of oysters have brown, black, or orange margins. The margins regulate the flow of water entering through the shell valves when they are open. The edges of the mantle have many tentacles that perform a variety of functions, including straining the water to keep coarse particles from entering the delicate filtering system of the oysters and warning the oyster of chemical changes in the surrounding water. The tentacles are also sensitive to light, enabling the oyster to detect an approaching predator when its shadow falls across the mollusk.

Near the center of the body is a large adductor muscle, attached to both valves, which controls the closing and opening of the shell.

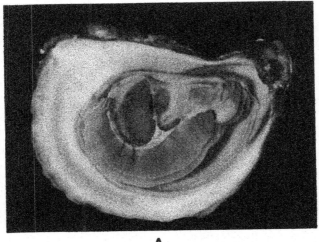

A

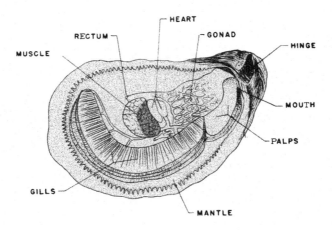

B

Figure 2.--Photograph (A) and diagram (B) showing main features of the anatomy of the oyster lying in its left shell and having the right shell and right half of the mantle removed. Detailed description of anatomy is given in text.

4

When it contracts the shell closes, but upon its relaxation the elastic properties of the ligament located at the hinge force the valves apart. Even superficial examination of this muscle shows that it consists of two different parts (fig. 2): a semitranslucent part near the hinge, and an opaque, white part surrounding the first part. Microscopic examination of the tissues composing these two parts of the muscle also shows that their fibrils differ in size and structure.

Physiological studies of the muscle have shown clearly that the two parts perform distinct functions. The translucent area can contract very quickly but, on the other hand, it soon tires and is compelled to relax. Therefore, it could be called appropriately the quick muscle. The opaque portion, however, contracts relatively slowly but can remain contracted for long periods while expending little energy. Furthermore--and this is important--when the opaque portion of the adductor muscle contracts, it becomes locked as if by a ratchet; hence, it is known as the catch muscle. This long-sustained contraction may be caused by a continuous discharge of nerve impulses. The ability of the oysters to stay closed for several days or even weeks is extremely useful because it helps them survive unfavorable conditions, such as freshets or temporary pollution of the water.

Shell movements of an oyster caused by contraction or relaxation of the adductor muscle can be recorded easily by several means. One method consists of recording the shell motions on a kymograph (an apparatus that has a recording pen and paper mounted on a drum which rotates at a constant speed). Studies of shell movements are now commonly used to observe behavior of oysters under normal and abnormal conditions. For example, records obtained on shell movements of oysters exposed to different concentrations of pollutants in sea water showed the response of the oysters to their presence and the level of tolerance of these substances. Usually, in the presence of pollutants or during unfavorable physical changes in the environment, the rhythm of opening and closing of the shells is changed, and movements of the valves become irregular. If the concentrations of harmful substances become too strong, the shells of the oysters remain closed.

The main body of the oyster lies between two folds of the mantle which is attached to it. Under the mantle at the anterior end of the body, nearest to the hinge, are four thin lips, or palps (fig. 2). After the lips are four rows of sickle-shaped organs known as the gills, arranged one below the other like pages in a book. The gills extend almost from the mouth, which is hidden under the palps, to about two-thirds of the distance around the body. At their end the mantle lobes of the two sides are united, and this union divides the mantle

cavity into a large, inhalant chamber containing the gills, and a much smaller, or exhalant, chamber. The water enters the inhalant chamber and leaves the oyster body by the exhalant chamber. Therefore, the gills form a complete partition dividing the mantle cavity. The only passage from one chamber to another is through the fine interstices or openings between the cross-connected filaments of the gills.

As Yonge indicates in his book, the precise region where the water enters the body of the oyster is controlled by the margin of the mantle. Usually this margin separates only in the central region of the inhalant chamber so as to concentrate the inflowing water and increase its speed.

The gills of an oyster, and of many other bivalves, are complex organs principally concerned with respiration and feeding. In general, these gills may be compared to a fine sieve. The openings of the sieve, called ostia, are surrounded by structures resembling microscopic whips, or cilia, which beat inward in an orderly manner, producing a current of water that passes through the pipelike structures inside the gills and is finally discharged through the exhalant or cloacal chamber.

Surfaces of the gills also are covered with cilia arranged in definite rows. They lash continuously, creating a current of water, and push the microscopic algae and other small particles toward the edges of the gills. The material that settles on the gill surfaces is entangled in mucus secreted by special cells of the gills and is carried gradually toward the mouth. Before the material is swallowed it is sorted, first on the gills themselves but especially in the region of the palps. A portion of the material may be rejected in the form of pseudofeces, which consist of mucus secretions produced by oyster gills in which small marine algae, detritus, and particles of turbidity-causing materials are embedded. When food organisms or turbidity-creating particles, such as silt, are too numerous, the oyster may reject as pseudofeces the largest portion of material collected on its gills. In general, oysters, like many other bivalves, feed most effectively in relatively clear water.

Scientists have found that the rate of water pumping is affected by several factors, including temperature, salinity, turbidity, and the presence in the water of various chemical substances. Most Eastern oysters stop feeding and hibernate when the water temperature decreases to about 41° F. Some adult Long Island Sound oysters, however, pump water at temperatures as low as 34° F. but, as a rule, the rate of pumping remains low as long as the temperature of the water is below 46° or 47° F. Within the range of about 47° to 61° F. the rate rapidly increases, but between 61° and about 82° F. the increase is comparatively slow. Between 83° and 90° F. a further increase in the pumping rate occurs, and it is within this range that the maximum average pumping rate of about 3-1/2 gallons per hour per oyster was recorded. Beyond 93° F., however, oysters begin to show a marked decrease in pumping rate, and their shell movements become abnormal.

The maximum rate of pumping for an individual oyster was registered at a little less than 10 gallons per hour. For shorter periods of 5-15 minutes the rate of pumping of the same oyster exceeded 10-1/2 gallons per hour. This oyster was only about 4 inches long; it is probable that larger oysters can pump even larger quantities of water per given unit of time.

As mentioned before, oysters feed most efficiently when the surrounding water is relatively warm. Under favorable conditions the oyster keeps its shell open about 22 hours during a 24-hour period. There is no correlation between opening and closing of the shell and the time of day.

The pumping rate of oysters kept at temperatures below 40° F. and then quickly changed to a temperature of about 65°-68° F. was virtually the same as the control oysters, thus indicating that the oysters respond and adjust to such radical changes. Oysters, therefore, are physiologically well adapted to rapid changes which they sometimes encounter in nature, for example, when living on tidal flats, where at autumn low tides the night air may cool the water of the small pools containing the oysters almost to freezing, while during the day the incoming tide may cover the same oysters with much warmer water.

Turbidity caused by various substances, including natural silt, also may affect the rate of pumping. Although very small quantities of silt sometimes stimulate the normal pumping activities of oysters, heavier concentrations significantly reduce the rate of water pumping and strongly affect the shell movements. Long Island Sound oysters living in relatively clear water reduced the pumping rate to about 68 percent of normal after being exposed to a concentration of silty water (one-fiftieth of an ounce of silt per quart of water). Greater concentrations of silt more strongly affected shell movements and rate of water pumping. In turbid waters the oysters discharged large quantities of pseudofeces containing silt. Shell movements of oysters kept in turbid waters were clearly associated with frequent ejection of large quantities of silt and mucus accumulated on the gills and palps. The size and shape of turbidity-producing particles are important. Different turbidity-creating substances, when present in the water in the same concentrations, affect experimental animals in different degrees.

The oyster eats plankton, which consists of microscopic plants and animals living in the water. According to some authorities, organic detritus, the product of disintegrating plants and animals, may also contribute to the oyster diet. Many kinds of marine bacteria are also ingested and possibly some may be digested. No definite experimental proof exists that oysters can absorb nutriments directly from sea water. Oysters gather food with their gills, and in filtering the water through the gills the oyster retains many microscopic organisms, although some small, elongated forms without appendages may escape. Only 10-50 percent of the bacteria present in sea water are detained by the gills.

The food particles caught on the gill surfaces are embedded in the mucus and are pushed along the upper or lower edges of the gills to the palps, which either may direct food into the mouth or reject it. Unwanted material is expelled by a sudden closing of the valves.

The mouth of the oyster lies between the palps and opens into the esophagus, which leads to the stomach. The stomach opens into the intestine, a long, coiled tube ending in the vent or anus. The stomach is surrounded by a brown digestive gland, which in fat oysters is obscured by the mantle and by gonad material, but in poor oysters, especially after the spawning period, shows clearly through the surrounding tissue. This dark-colored mass of tissue is sometimes misnamed the liver, but usually biologists call it the digestive diverticula. A series of ducts unites this organ with the stomach.

As has been described by many authors and so well summarized by Yonge in his book, the digestive system of the oyster possesses a most remarkable structure. Called the crystalline style, it is a gelatinous rod and occurs only in other bivalves and in certain snails. The head of the style projects from the elongated style sac where it is formed and extends across the stomach cavity, pressing against an area known as a gastric shield, which is the only area in the stomach not covered by cilia. The crystalline style rotates continuously. This rotation assists in mixing the food, aids digestion, and also brings particles of food in closer contact with the stomach walls. The crystalline style, incidentally, is the only known rotating part in any animal.

The style, along with several other organs of the oyster, including the digestive diverticula and even blood cells, or leucocytes, is concerned with digestion. It is made of protein produced in a long style-sac which is lined with numerous cilia that rotate the style. Normally, under healthy conditions the style is added to continuously by formation of new material. As a result, the style is pushed forward while it is rotated, and its head, directed against the gastric shield of the stomach, dissolves continuously. The dissolution of the style releases digestive enzymes that convert starches and some celluloses into the simple sugar called glucose and also probably break down fats. These enzymes are needed because many plant cells, which comprise the major items of oyster food, contain large quantities of starch. The material of the dissolved style also helps to lower stomach acidity to provide optimal conditions for digestion. Thus, the style helps mix the food particles in the stomach and also provides a continuous supply of digestive enzymes.

Since the particles which enter the tubes of the digestive diverticula from the stomach are small, they can be ingested into the cells and go through the so-called intracellular process of digestion, in contrast to the extracellular type of digestion taking place in the stomach proper. Both proteins and fats can be digested in this manner. Undigested portions of the particles are discharged from the cells. They are finally forced to enter the intestinal groove and are eliminated in the normal way.

Blood cells of the oysters also participate in digestion. These phagocytic cells pass through the stomach walls into the stomach, where they engulf small particles such as diatoms. Digestion involves the gradual breakdown of ingested particles, after which the blood cells move back through the stomach lining into the blood stream.

Part of the absorbed food is stored in the oyster body. Oysters are considered "fat" when their meats are large and plump. This condition, however, may be caused either by development of spawn or by accumulation of reserve food material. A plump condition during breeding time results from the increased gonads; the fatness of the oyster after the breeding season, when the oyster feeds actively, to the reserve food. The stored material contains many substances, but it consists mainly of animal starch (glycogen), which is found in the large cells of connective tissue in most parts of the body.

In oysters of Long Island Sound, accumulation of glycogen commences sometimes as early as August and continues until the start of winter hibernation, which is usually early in December when the water temperature decreases to about 41° F. In the warmer waters of Florida, Louisiana, and Texas, fattening may not begin until December. In all cases, the stored food reserves are normally used during the following spring and summer when the oyster grows and forms its reproductive cells.

The quality of oyster meats depends principally upon its solids and amount of stored glycogen. Good quality oysters usually contain between 18 and 20 percent solids; poor ones may contain less than 10 percent. Normally, a high solids content is accompanied by correspondingly high glycogen storage. "Fat"

oysters are normally white, and their meats fill the shell cavity.

Oyster meat contains large quantities of nutritive substances that people need for a balanced diet. It is high in copper and iron, which are needed for proper composition of human blood, and because of this, oysters are often prescribed for patients with anemia. Oyster meat also contains iodine necessary for normal activities of the thyroid gland. Proteins in oyster meat are especially high in nutritive value, and the carbohydrates, in the form of glycogen, are readily digested and assimilated. Phosphorus and calcium are present in relatively high quantities. Oyster meat also contains most essential vitamins.

The condition of oyster meats depends on several factors, such as location of oyster beds, quantity and quality of food present in surrounding waters, salinity of the water, and time of year.

The provision that oysters should not be eaten in months that do not contain the letter "r", while valid for the flat European oyster which incubates its young during summer, does not apply to American oysters, which can be and are eaten at any time of the year. Because they are highly perishable, however, their transportation and storage before refrigeration was widespread created considerable difficulties during warm weather. Furthermore, oysters that spawn late in summer become watery and unattractive in appearance and are not as acceptable as during the cold season, when the quantity of stored materials is highest. Nevertheless, there are localities in several States where oysters of fairly good quality may be available during the entire year.

The circulatory system of the oyster carries body fluids from one part of the animal to another. The heart is located above the adductor muscle in the pericardial chamber. It consists of a single ventricle and a pair of contractile auricles, one on each side. The auricles collect blood largely from the gills and push it to the ventricle, which drives it by rhythmical contractions into the anterior and posterior aortas. The posterior aorta is short and supplies only the adductor muscle and the rectal region. The rest of the body is supplied by the anterior aorta, which divides into a series of smaller blood vessels. These vessels open into the so-called blood sinuses, where the blood flows and washes the different organs coming in contact with it. The used blood, now low in oxygen, collects in the veins and is carried into the gills or the organs of excretion, the kidneys, sometimes called the organs of Bojanus. The kidneys, which purify the blood, are located by the adductor muscle and consist of two convoluted tubes connected internally with the pericardium and externally with the exhalant chamber.

With the assistance of so-called accessory hearts the blood from the kidneys is pumped into the vessels of the mantle, and eventually this blood, together with blood from the gills, is returned to the heart by way of the auricles. The blood cells are colorless and contain neither the red hemoglobin that is found in blood cells of higher animals, nor hemocyanin, a blue-colored substance found in other mollusks, such as some snails, squids, and octopuses.

The nervous system of the oyster is comparatively simple. It consists of two knots, or ganglia, of nervous tissue situated near the mouth, from which two nerves pass to another pair of ganglia lying under the adductor muscle. From these two pairs of ganglia small nerves extend to various parts and organs of the body. Oysters, like other mollusks, do not possess the type of centralized nervous system which is characteristic of vertebrates.

GROWTH

Oyster growth varies considerably with size, temperature, quantity and quality of food, and seasons of the year. As a rule, growth is more rapid in warm waters, such as the Gulf of Mexico, where a marketable size of 3-1/2 inches may be reached on some beds in 2 years. In northern waters, such as Long Island Sound, characterized by shorter summers and generally lower water temperatures, oysters reach this size in 4 or 5 years. The average size of a Long Island Sound oyster at the end of the first, second, third, fourth, and fifth growing periods is 3/4, 2-1/4, 3, 3-1/2, and 4 inches, respectively. Specimens measuring more than 14 inches long have been found on old natural beds.

The exact maximum age attained by oysters cannot be stated definitely, but the number of layers composing the shells of some unusually large individuals indicates that they may reach an age of 40 years.

REPRODUCTION

Gonad Development and Spawning

Oyster reproductive organs, or gonads, consist of a mass of tissue made of microscopic tubules, sex cells, and connective tissue which envelops the stomach, digestive diverticula, and the fold of the intestine, and, during the period of ripeness, constitutes a significant portion of the entire oyster body (fig. 2). The gonads become larger and thicker as their eggs and sperm mature. When ripe, the gonadal layer in the region of the stomach of an oyster about 4-1/2 inches long may be as thick as a quarter of an inch.

The Eastern oyster exhibits alternative sexuality; i.e., adults function seasonally as separate sexes. In this respect, it is like the Japanese oyster, the Australian oyster, Crassostrea commercialis, and the Portuguese oyster, C. angulata. It differs, however, from another group typified by the European oyster, which has a series of alternating male and female phases throughout its life.

During the first spawning season of the Eastern oyster, most young individuals function as males. This condition is known as protandry. Nevertheless, the gonads of young oysters may show all gradations from true males, in which no female cells can be found, to those that develop directly into ovaries containing growing eggs. After the second spawning season the numbers of individuals of each sex are almost equal.

The adults function seasonally as separate sexes, but, nevertheless, the sex of an oyster is generally unstable and, therefore, a change of sex from male to female and vice versa sometimes occurs. This change usually takes place in the intervals between the two spawning seasons. In experiments with Long Island Sound oysters, about 9.7 percent reversed their sex. The percentage of reversals was considerably higher among the females (13 percent) than among the males (8 percent). Functional hermaphrodites are found among adult oysters, but, as a rule, they are uncommon, constituting less than 1 percent of the adult population.

Gonads of Eastern oysters change markedly during the year. Such changes may vary considerably from one geographical area of the Atlantic or Gulf coast to another, but, nevertheless, they correspond to definite seasons of the year and, in general, markedly affect the physiological behavior of the oysters and the chemical composition of their meats.

During the cold season, the gonad follicles of oysters of northern waters, such as those of Long Island Sound, are small and contain sex cells only in early stages of development. With the end of hibernation, however, as soon as the temperature of the surrounding water increases, the gonads develop at first slowly and later rapidly, and by the end of June many oysters are already ripe (fig. 3).

Oysters begin to spawn as soon as the water temperature is sufficiently high. Gonads of partially spawned oysters are characterized by contraction of the follicles, invasion by phagocytes, and rapid increase of connective tissue cells that contain much glycogen. Resorption of the gonads is completed in October. After October they enter a brief indifferent stage during which the sexes are almost indistinguishable. It is assumed that sex reversal, if it occurs, takes place during this period when the sex is least defined. At the end of the indifferent phase, gonad development begins again but it is soon stopped by the low temperatures of approaching winter. In the spring, gonad development is resumed.

8

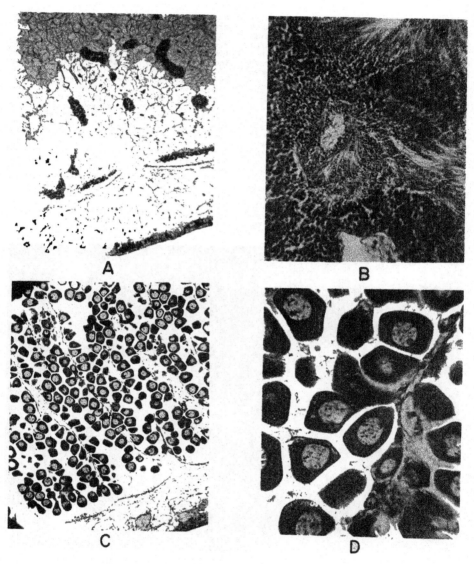

Figure 3.--Portions of gonads of oysters in different stages of development.
 A. Gonad in early spring before active development of sex cells begins.
 B. Gonad of ripe male containing large number of spermatozoa (center).
 C. Gonad of ripe female containing numerous eggs. Low magnification.
 D. Gonad of ripe female showing highly magnified eggs.

Oysters are highly prolific. In one experiment, a female 4-1/2 inches long released about 70 million eggs in a single spawning. Larger oysters can develop and discharge more eggs. A female or male may spawn a number of times. Most ripe oysters of both sexes spawned after being stimulated to spawn on 5 consecutive days.

There was no significant difference in the average number of eggs released during a season whether the oysters were spawned at 3-, 5-, or 7-day intervals until they were completely spent. Female oysters with larger numbers of eggs at the beginning of the season spawned more frequently than females with fewer numbers.

Until several years ago it was unknown whether there is an age in the life of oysters when they produce the best, most viable sexual products. Recently, experiments with oysters ranging in age from 2 to 30 or 40 years (fig. 4) showed conclusively that there is no significant difference in the quality of spawn developed by individuals of different ages or sizes and, therefore, mature oysters of all age groups may be used safely as spawners.

The oysters of the oldest group, some over 9 inches long and 4 inches wide, responded to the spawning stimuli somewhat faster than individuals of the youngest group. There was no significant difference in the percentage of fertile eggs because almost all eggs of all age groups were fertilized. Moreover, the percentage of eggs developing to the straight-hinge larval stage showed no variations that could be ascribed to size or age of parent oysters. Finally, no consistent difference was found either in the size of the early straight-hinge larvae developed from eggs discharged by oysters of different age groups or in survival and growth rate of these larvae.

Of special interest was the observation that the sexes of the oldest oysters were about evenly divided, thus contradicting the old, often-expressed conception that females usually predominate among the oldest and largest oysters.

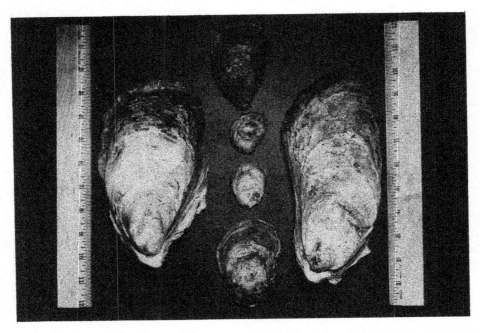

Figure 4.--Eastern oysters of different sizes and ages. The two smallest oysters are about 2 years old and the largest are between 30 and 40 years of age. Regardless of the difference in sizes and ages of the oysters their spawn showed equal viability.

The length of the spawning period of oysters depends upon the climate. In New England waters it extends on the average from 2-1/2 to 3 months, while in Florida waters oysters with ripe eggs or spermatozoa can be found most of the year. Therefore, until recently, in New England waters and similar areas experiments on larvae of oysters and most other bivalves were confined exclusively to the short period of natural propagation. However, biologists of the Fish and Wildlife Service, Bureau of Commercial Fisheries, recently found that, by using proper methods, normal development of gonads and spawning can be induced during late fall, winter, and spring (fig. 5).

Conditioning oysters to develop mature spawn during winter is relatively simple. Place the mollusks in warm water and then gradually increase the temperature to the desired level. Toward spring, instead of slowly conditioning the mollusks by gradually increasing the water temperature, the oysters can be placed directly in water of about 70° F. and kept there until they become ripe, which normally

takes from 3 to 5 weeks. At higher temperatures they will ripen more rapidly. Upon ripening, the oysters can be induced to spawn in the laboratory by quickly raising the temperature to about 77° F. and adding a suspension of eggs or spermatozoa to the water (fig. 6). The spawn obtained in this way is no less viable than that released by oysters in the usual manner during summer.

Fertilizable eggs can be obtained by stripping ripe females, but since oysters spawn so readily in response to chemical and thermal stimulation, this method is seldom needed. Moreover, among the eggs obtained by induced spawning there are always considerably fewer abnormal ones than among eggs obtained by stripping.

In early fall, when Long Island Sound oysters cannot be conditioned because they have not recovered from summer spawning, ripe spawn can be obtained from individuals placed, in spring, in water sufficiently warm to allow the eggs to develop to maturity but not high enough to permit spawning. This has been done successfully with Long Island Sound

Figure 5.--Milford Biological Laboratory, one of the centers where extensive research on oysters and other commercial mollusks is conducted by the Bureau of Commercial Fisheries.

Figure 6.--Biologist inducing out-of-season spawning of oysters and clams under laboratory conditions.

oysters by shipping them, early in May, to Boothbay Harbor, Maine, where the water temperature is considerably cooler than in Long Island Sound. Oysters so kept can be induced to spawn any time between mid-August and late November, thus providing normal sex products which then are unobtainable from Long Island Sound oysters.

By use of the two above-described methods of advancing and delaying gonad development, spawning of the Eastern oyster can be induced any time of the year. As a result of these discoveries, much more can now be accomplished in the field of spawning and propagating oysters and some other species of commercial mollusks.

The conditioning methods described above are not equally successful with all groups of the Eastern oyster. This is probably because populations of this species are not genetically alike, but consist of different physiological races. Some experiments in this field strongly support this assumption by demonstrating that, even though all these oysters belong to the same species, the temperature requirements for gonad development and spawning of the

northern populations are definitely lower than for the southern group. In some of these experiments it was possible in winter to induce spawning of 50 percent of Long Island Sound oysters after only 18 days of conditioning at about 71° F., while 78 days were needed to achieve the same results with New Jersey oysters. Oysters of the more southern groups, kept under conditions identical with those applied to northern oysters, failed, as a rule, to produce 50 percent spawners.

Eggs and Larvae

Fertilized oyster eggs vary in diameter from about 45 to 62 microns, but the majority measure between 50 and 55 microns. A micron is one twenty-five thousandth of an inch and is designated by the Greek letter "μ". The size of an egg is not influenced by the size of the mother oyster; for example, eggs discharged by females measuring over 9 inches long averaged 50.4 μ, while eggs from younger and smaller females only 3 to 4 inches long averaged 51 μ.

12

If kept under favorable conditions, 90 to 95 percent of fertilized eggs will develop to shelled veliger stage, often called the "straight-hinge" stage. The smallest normal straight-hinge larvae measure about 68μ long by 55μ wide (or deep). At about 72° F. the larvae will attain 75 by 67μ within 48 hours (fig. 7). At this stage some larvae already begin to take in food composed of microscopic water organisms, principally plants.

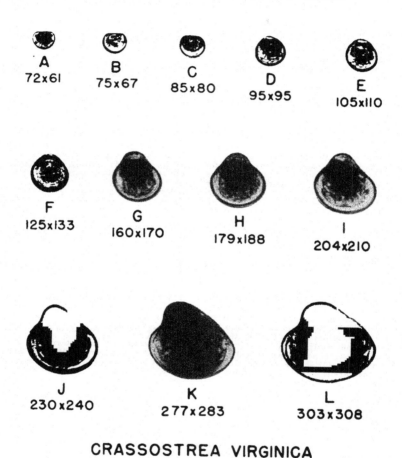

A
72x61

B
75x67

C
85x80

D
95x95

E
105x110

F
125x133

G
160x170

H
179x188

I
204x210

J
230x240

K
277x283

L
303x308

CRASSOSTREA VIRGINICA

X 112

Figure 7.--Photomicrographs showing several stages of development of oyster larvae from straight-hinge to metamorphosis. The figures below each larva give its length and width in microns. (A micron (μ) is 0.001 mm.) Magnification is about 112 times.

13

As the larvae grow, their length, measured parallel to the hinge line, continues to be about 5 to 10μ more than their width, extending perpendicularly from the hinge to the ventral edge of the shell. This condition persists until larvae reach about 85 to 80μ. At 95 to 100μ the length and width are about equal, but from then on the increase in width is more rapid than in length. When the length reaches 125 to 130μ, the width already exceeds it by about 8μ, and this disparity remains until the larvae reach metamorphosis (when they change to a more oysterlike form).

A well-pronounced black spot, called an "eye", develops when the larvae are about 250-275μ long and remains throughout the rest of the free-swimming period. Most larvae undergo metamorphosis; i.e., begin to set, between 275 and 315μ although occasionally free-swimming larvae may be as long as 355μ. The larvae are highly active and remain in suspension throughout most of the free-swimming period. Large larvae, measuring about 200μ or more, tend to gather on the water surface in hatchery vessels and form small "rafts" that float just below the surface film. If disturbed, the larvae composing the "rafts" swim apart, but congregate again in a short time.

Growth of larvae is affected by many conditions but chiefly by food and temperature. Their effects will be discussed later. However, even though larval cultures sometimes originate from the same spawnings of the same parents and are grown under identical conditions in the same vessel, individual larvae may show a widely different rate of development and growth and, therefore, metamorphose at different times. For example, in one healthy culture of larvae kept at about 73° F. the first individuals began to set 18 days after fertilization. The intensity of setting gradually increased and remained heavy for 17 days, but some larvae continued to swim another 10 days before metamorphosing. Thus, setting of this presumably homogeneous culture of larvae continued for 27 days. Occasionally, however, the larvae in some uncrowded cultures may be of relatively uniform sizes (fig. 8).

Contrary to the old opinion, oyster eggs and larvae, if protected against disease-causing organisms and toxic substances, are rather hardy and are able to withstand sharp changes in their environment. For example, if laboratory-grown larvae kept at a temperature of 72° F. are placed in the refrigerator for 6 hours and then returned to room temperature, most will recover and continue to develop normally. It seems unlikely, therefore, that ordinary short-term temperature variations of a few degrees, occurring in natural waters, can be responsible for heavy mortality of larvae.

It will be shown later in this article that eggs and, especially, larvae of oysters can also endure significant changes in salinity and turbidity. They can withstand strong mechanical disturbances, such as vigorous water motions created by winds of hurricane proportions, without ill effects. On the other hand, eggs and larvae show sensitivity to even traces of certain chemicals present in sea water. Some of these are natural substances released from the bottom soil or produced by plants and animals living in the sea. Other substances, such as insecticides, weedicides, oils, organic solvents, and detergents, may strongly affect eggs and larvae even if the substances are in minute concentrations. For example, of commonly used insecticides, DDT is one of the most toxic to oyster larvae because, even at a concentration of 0.05 part per million (p.p.m.), it killed nearly all of these organisms. However, another common insecticide, Lindane, even when present in sea water at a concentration of 10.0 p.p.m., caused no appreciable mortality among larvae of the Eastern hard shell clam, Mercenaria mercenaria. Effects of each insecticide, therefore, should be evaluated separately.

Metabolites, substances released into sea water by many aquatic micro-organisms, especially the one-celled organisms called dinoflagellates, seriously affect oyster eggs, larvae, and adults. Dinoflagellates are the forms responsible for the so-called "red tide" in Florida and in other States, which sometimes kills not only shellfish but finfish. Experiments at one of the Bureau's biological laboratories have shown that fertilized oyster eggs placed in sea water containing a large number of dinoflagellates became unfavorably affected, and most of the eggs were unable to develop into normal shelled larvae.

Effects of Temperature on Eggs and Larvae

Experiments to determine temperature limits for development of oyster eggs showed that at 60° F. none of the eggs reached normal straight-hinge stage, although a few developed as far as early shelled larvae. At 64° F., however, about 97 percent of the eggs developed to fully formed straight-hinge stage. In experiments at 86° F., a temperature found in southern waters, most fertilized eggs developed into normal straight-hinge larvae. At 92° F., however, about half of the eggs reached straight-hinge stage, and many were abnormal.

At all favorable temperatures, growth of oyster larvae depends, to a large extent, upon the food available. Thus, when kept at the same temperature, larvae given relatively poor food, such as microscopic green algae called Chlorella, grew less rapidly than larvae given better food. Nevertheless, even when

Figure 8.--Group of larvae of relatively uniform size, about 175 μ, from a well-kept culture. Note protruding umbo of the larval shells, which is one of the characteristics that distinguish advanced larvae of the genus Crassostrea from the larvae of most other genera of bivalves.

fed Chlorella, the larval growth within 68° to 87° F. increased progressively with each increase in temperature. When given good food at the latter temperature the larvae began to metamorphose 10-12 days after fertilization (fig. 9).

If oyster larvae are reared almost to the stage of metamorphosis at 80° F. and then transferred to water temperatures of 77°, 73°, 71°, 68°, and 64° F., the intensity and duration of setting of these different groups are profoundly affected by water temperature. For example, all larvae transferred to 77° F. underwent metamorphosis and set within 8 to 10 days. Those kept at 73° F. required from 12 to 16 days to complete the setting. At 68° F. the slowest growing larvae required 20 days to set. At 64° F. the setting period was even more prolonged. In some instances fully mature oyster larvae grown at optimum temperature can set even if transferred to a temperature as low as about 55° F. In general, within a certain temperature range the number of spat obtained from a given number of mature

larvae decreases as the temperature to which they are transferred is decreased. This reduction, however, is not as drastic as might be anticipated. Even when mature larvae are transferred to a low temperature of about 64° F. about half as many larvae as at 77° F may eventually metamorphose. All this knowledge obtained by laboratory experiments helps us to understand the behavior of larvae in nature and assists in the management of our oyster resources.

Effects of Salinity on Eggs and Larvae

Experiments demonstrated that the optimum salinity for development of eggs of Long Island Sound oysters is about 22.5 parts per thousand (p.p.t.). Nevertheless, some normal larvae originating from these eggs developed in salinity as low as 15.0 p.p.t. and as high as 35.0 p.p.t. At salinities lower than 22.5 p.p.t., however, the percentage of eggs developing to straight-hinge stage steadily decreased until at 15.0 p.p.t. only 50 to 60 percent of the eggs

15

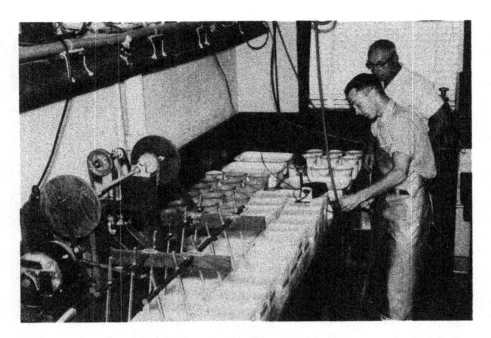

Figure 9.--Biologists conducting experiments with oyster larvae to determine their food requirements and the optimal ranges of temperature, salinity, and other factors of the environment for larval development and growth.

reached this stage. At 12.0 p.p.t. practically no eggs of Long Island Sound oysters developed into normal shelled larvae.

The optimum salinity for shelled larvae developed from eggs of oysters grown in Long Island Sound at a salinity of 27.0 p.p.t. was about 17.5 p.p.t. Good larval growth was also recorded at 15.0 p.p.t., while at 12.5 p.p.t. growth was appreciably slower, although some larvae grew to metamorphosis even at this salinity. At 10.0 p.p.t. the growth practically stopped. This was true, however, only in relation to straight-hinge larvae, because the older the larvae became, the better they withstood such low salinity, although growth in all size groups was significantly reduced.

Another experiment used Maryland oysters that grew and developed gonads in the upper part of Chesapeake Bay, where the salinity at the time of collection was only 8.7 p.p.t. Spawnings were induced in salinities of 7.5, 10.0, and 15.0 p.p.t. Some discharged eggs developed into normal larvae at 10.0 and even 7.5 p.p.t. although, in the latter, abnormally small individuals were quite common. In general, the optimum salinity for development of

eggs of this group of oysters was between 12.0 and 15.0 p.p.t., while a salinity of about 22.0 p.p.t. was the upper limit for normal development.

Effects of Turbidity on Eggs and Larvae

Until recently virtually nothing was known of the ability of oyster eggs to develop and larvae to survive in turbid waters. Recent experiments with oysters native to Long Island Sound demonstrated that natural silt is harmful to eggs and larvae. Thus, in a concentration of 0.25 gram of silt per liter of water (g./l.), only 73 percent of the eggs survive. In a concentration of 0.5 g./l., only about 31 percent of eggs survive and continue to develop. In stronger concentrations of 1 or 2 g./l. practically no eggs develop to straight-hinge larval stage.

Larvae were also significantly affected when the concentration of silt was 0.75 g./l. At concentrations of 1.5 g./l., or higher, growth was negligible and no oyster larvae survived to metamorphosis in concentrations of 3 and 4 g./l. However, some larvae may survive,

16

without showing an increase in size, for 14 days in concentrations of silt as high as 2 g./l. Recent experiments also have suggested that the sizes and shapes of turbidity-creating particles are important in the degree of damage they cause eggs and larvae. These studies are now in progress.

Food and Feeding of Larvae

Oyster larvae, like other living organisms, must feed to survive and grow. Larvae begin to take particulate food soon after they reach straight-hinge stage. Since the diameter of their gullet is then only about 9 μ, particles they can digest are correspondingly small.

Most food organisms consumed by the larvae are members of a large group of microscopic marine algae. Not all species of algae, however, are of equal value. The thickness of algal cell walls and the degree of toxicity of metabolic products that algae produce are important factors in determining their quality as larval food. Critical evaluation of a number of micro-organisms has shown that so-called naked flagellates, such as Isochrysis galbana and Monochrysis lutheri, produce few, if any, external metabolites which may unfavorably affect larvae.

The food value of micro-organisms depends upon how completely they meet the requirements of larvae. Sometimes a mixture of the two above-mentioned flagellates, together with two other naked flagellates, Platymonas sp. and Dunaliella euchlora, induced more rapid growth of larvae than equal quantities of these food organisms given separately. All these forms induce better growth than forms with thick cell walls, such as several species of Chlorella.

Experiments are now in progress to find dried foods that may be used successfully to feed molluscan larvae. So far, these experiments have shown that some dried algae can be used for growing to metamorphosis larvae of some bivalves, such as the common hard clam, Mercenaria mercenaria. Unfortunately, even flagellates, such as I. galbana and M. lutheri, are not good enough, when dried, to maintain oyster larvae. Nevertheless, it is hoped that this problem will be solved soon.

Metamorphosis, or Setting, of Larvae

At the close of the free-swimming period, when a larva is about 300 μ, or roughly one seventy-fifth of an inch long, it drops to the bottom to metamorphose. After descending, it crawls for some time and may even swim away to a new area until it finds a firm surface free of dirt or silt, to which it attaches itself by its left valve. Before, during, and immediately after metamorphosis, radical changes occur in the anatomy of the setting organism, such as disappearance of the velum and foot and development of a primitive set of gills. Recently metamorphosed oysters are usually called "set" or "spat" (fig. 10).

In many bodies of water the spat attaches itself at all depths from the surface to the bottom. In Long Island Sound, oysters set even at a depth of 100 feet. In other areas of the Atlantic and Gulf of Mexico coasts setting may be confined to the zone between tidal levels or only near the bottom, regardless of depth.

Experiments have shown that during setting oyster larvae prefer certain surfaces or materials (fig. 11). If old, clean oyster shells and artificial collectors made of plastic, cement, glass, or resins are placed in the same basin with mature, ready-to-set larvae, the oyster shells will collect considerably more spat, per square foot of surface, than artificial collectors.

Diseases of Larvae

It is quite difficult to establish whether, in nature, oyster larvae suffer from various diseases. In the laboratory, however, even under the best conditions, heavy larval mortalities, which could be ascribed to diseases or parasites, often occur at all ages. In 1954 the Bureau's biologists discovered that a fungus, Sirolpidium zoophthorum, may be responsible for epizootic mortalities in cultures of bivalve larvae, including oysters. Still more recently, it was demonstrated that certain bacteria produce toxins that can retard growth or even kill larvae. Some bacteria that kill molluscan larvae were tentatively identified as belonging to the genus Vibrio or Pseudomonas.

High temperatures usually favor bacterial growth. It also has been shown that under hatchery conditions bacterial contamination of algal food cultures sometimes causes a sharp decrease in growth of oyster larvae without causing extensive mortality. Soon after the discovery of pathogenicity of fungi and bacteria, biologists began testing a number of antibiotics and fungicides to prevent and control larval diseases without seriously affecting survival and growth of larvae. A number of such substances now are used in proper concentrations with considerable success in pilot oyster and clam hatcheries.

OYSTER ENEMIES

Diseases and Parasites

Oysters in all stages of development are susceptible to attack by numerous enemies. The oyster eggs, early embryos, and larvae are eaten in large numbers by protozoans, ctenophores, jellyfish, hydroids, worms, bivalves, barnacles, numerous larval and adult

17

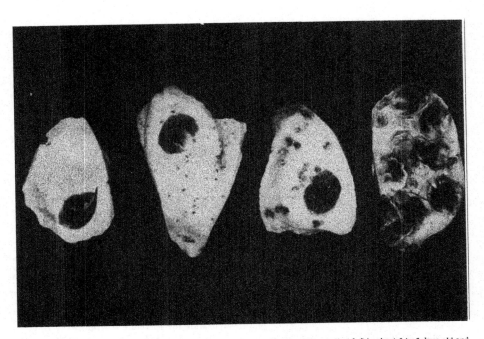

Figure 10.--Oyster set at different ages and sizes. The smallest set seen on the shell at left is about 3 to 5 days old and measures only about 1 mm. in length. The set on the shell at the right was photographed at the end of summer and measures about 3/4-inch. The two shells in the middle show intermediate sizes.

crustaceans, especially crabs, many varieties of larval and adult fish, ascidians, and others. It is probable that under natural conditions oyster larvae are parasitized by fungi, such as S. zoophthorum, and are infected by certain bacteria, such as Vibrio and Pseudomonas. Perhaps viruses also kill oyster larvae and adult oysters.

Oysters are parasitized with many micro-organisms, some of which are responsible for extensive mortalities. One of the most destructive cases of this nature is the so-called Malpeque Bay disease, which struck oysters of Prince Edward Island, Canada, around 1915. Oysters affected by this disease become thin and emaciated and sometimes have yellowish pustules and abscesses. It took about 13 years for the oyster beds of Malpeque Bay to recover. Apparently, local oysters at that time became largely immune to this disease, but oysters originating in other locations, if transplanted to areas affected with Malpeque Bay disease, became infected. It is still uncertain what micro-organisms are responsible for the disease, and some authorities think that the

Malpeque Bay disease may be two or even three diseases.

Another widespread disease is caused by a highly pathogenic fungus, Dermocystidium marinum, which affects oysters from Delaware Bay to Mexico. The range is not continuous because certain areas, such as the Virginia seaside, appear free of this organism. In the northern range of its distribution, D. marinum disappears from oysters during cold seasons but reappears in summer to cause new losses. In the Gulf of Mexico, where water temperatures even in winter are relatively mild, considerable mortality may occur even during the coldest part of the year. Low salinities and strong tidal action usually decrease the strength of the epidemics.

In 1957 a heavy mortality of oysters occurred in Delaware Bay. In April and May from 35 to 85 percent of planted oysters died on some beds. The losses continued during 1958 and later were reported from Chesapeake Bay and other areas. This disease is now ascribed to a new organism called "MSX",

18

Figure 11.--Biologists experimenting with artificial collectors for oyster set to find a substitute for old shells, which are not easily available in some areas where oysters are cultivated. Artificial collectors may also be used in oyster hatcheries where mature larvae will be released to set on them after they are placed in small ponds or tanks similar to the one shown in photograph.

which may be a protozoan, probably belonging to the genus Haplosporidium.

Another organism, possibly a species of Haplosporidium similar to "MSX", is thought responsible for oyster mortalities along the Maryland-Virginia seaside. Therefore, it is possible that there are two species of Haplosporidium overlapping ecologically. For convenience's sake, the parasite from the Virginia seaside was given the name "SSO", signifying seaside organism.

Another protozoan often blamed for oyster mortality is a flagellate protozoan belonging to the genus Hexamita. In the past it was generally believed that Hexamita was a highly pathogenic organism. Now, however, biologists differ sharply in their opinions on the real role of this organism. One group believes it is a mild parasite; the second group thinks it is a parasite, not strong enough alone to destroy oysters, especially if they are in good physical condition, although it may contribute to mortality of weakened oysters after a long, cold winter or after exposure to adverse

chemical or physical conditions. Finally, the third group maintains that Hexamita is a highly dangerous parasite causing extensive mortality.

Still another protozoan parasite of the American oyster is Nematopsis ostrearum, a sporozoan protozoan which spends part of its life in crustaceans. Large numbers of Nematopsis spores are sometimes found in oyster tissues, especially in the gills and mantle. Two species of Nematopsis may affect oysters. The spores of one species, N. ostrearum, are found in almost all oyster tissue except the digestive system. The spores of another species, N. prytherchi, are located exclusively in oyster gills.

Small crabs, such as Panopeus, which carry one stage (heterogenic) of Nematopsis in their intestinal tracts, are very common on many oyster beds. The stages develop in the crab until gymnospores are formed. Eventually these spores are released by the crab, carried by currents, and some are ingested by oysters, infecting them. Crabs infect themselves by

19

feeding on dead oysters. It has never been conclusively demonstrated, however, that Nematopsis causes heavy oyster mortality.

A trematode worm, Bucephalus cuculus, inhabits the gonads of the oyster, sometimes causing complete castration. The parasite may invade other tissues and organs, eventually killing the oyster. A spirochaete bacterium, Crististira, occurs in the crystalline style of the oyster without causing any apparent damage to its host.

Small crabs, Pinnotheres ostreum, until recently considered as commensals and not parasites, are occasionally present in the oyster. Adults are found on the oyster gills facing the incoming current. They deprive the oyster of food it gathers and during this operation also injure its gills. Occasionally they feed on particles of oyster gills. The early stages of this crab, however, are truly parasitic. Sometimes many small crabs are found in a single oyster, not only on the gills but in other parts of the body. In general, infested oysters are in poorer condition than healthy ones, but there is no evidence of oyster mortality traceable exclusively to crab infestation.

Predators

Young and old oysters are attacked by many predators. Various marine snails, commonly called drills or conchs, are probably the most important and most widely distributed. Two species of drills, Urosalpinx cinerea (fig. 12) and Eupleura caudata, are found on most Atlantic coast oyster beds where the salinity of the water is high enough for their existence. Another oyster drill, Thais haemostoma, is considered by some authorities to be the most persistent oyster enemy in the Gulf of Mexico, especially in Alabama. While U. cinerea and E. caudata seldom are over 1-1/2 inches long, the southern drill, T. haemostoma, may reach a length of 3 inches. Another snaillike enemy of the oyster, Murex pomum, or apple Murex, is known to drill oysters in southern waters.

Drills attack oysters by boring small holes in their shells, then inserting a long proboscis, an organ equipped with a rasplike structure called a radula. With this instrument they scrape out pieces of oyster meat.

Several species of conchs, snails of considerable size, also feed on oysters. Among these, Busycon canaliculatum, or channel

Figure 12.--Group of common oyster drills, Urosalpinx cinerea, of the Atlantic coast.

conch, which attains a length of 6 inches, is quite numerous on certain oyster beds of the Atlantic coast. A conch opens an oyster by pushing the edge of its own shell between the oyster's valves. It then inserts its proboscis between the shells and eats the meat. Sometimes conchs chip the edges of the mollusk's shell until the proboscis can be inserted. In southern waters Busycon contrarium, or lightning whelk, destroys many oysters, while still another gastropod, the crown conch, Melongena corona, also may kill oysters although it is not considered a serious oyster enemy.

Predaceous snails can be controlled by chlorinated benzenes, such as orthodichlorobenzene, or a mixture of several such hydrocarbons known under the commercial name of Polystream. These chemicals are mixed with dry sand, broken oyster shells, or other inert material, such as clay, to carry them to the bottom and keep them there. The effectiveness of the treatment can be increased by incorporating in the chlorinated benzenes other chemicals, such as Sevin, an insecticide relatively nontoxic to mammals.

The treatment may consist of surrounding the shellfish beds with a belt of chemically treated sand, thus preventing entrance of enemies, or the sand may be spread over an entire infested area, seriously affecting or killing the drills and other undesirable gastropods (fig. 13). Because certain snails, such as the clam-killing snail, Polinices, are able to move under several inches of bottom soil, they may not be stopped by merely spreading the chemically treated sand. However, an effective three-dimensional barrier may be created by injecting the chemicals or plugs of treated sand into the bottom.

The methods described above were found quite effective when applied in nature. Studies are being made to ascertain if the method may be approved by the U.S. Public Health Service, which is interested in all instances in which chemicals come in contact with human food.

Still another group of small marine snails, belonging to the genus Odostomia, feeds on oysters. These snails attach themselves to the oyster body by an oral sucker, pierce the body wall with an organ resembling a stylet, and suck the oyster blood, perhaps also devouring some solid tissue.

The starfish, Asterias forbesi, found in large numbers in New England, especially in Long Island Sound, and sometimes in large concentrations in the lower part of Chesapeake Bay, is another important oyster enemy (fig. 14). It opens oysters by pulling the valves apart with its tube feet. The starfish mouth is

Figure 13.--Oyster conch, Busycon canaliculatum, in expanded condition caused by chemicals used for control of undesirable snails on oyster beds.

Figure 14.--Common starfish, *Asterias forbesi*, one of the most important enemies of oysters in Long Island Sound and other areas of the Atlantic coast.

very small and cannot take in large pieces of food. To overcome this difficulty, the starfish has developed an unique method of feeding. After the oyster shells are opened, the starfish extrudes its stomach and inserts it between the shells, digesting the soft oyster meat. As soon as the mollusk is eaten, the starfish withdraws its stomach with a set of retractor muscles. The starfish is very destructive, and a medium-sized individual may destroy up to seven 1-year-old oysters in a day (fig. 15). Older and stronger oysters are better equipped to withstand a starfish attack, but even a large oyster, if weak, may become easy prey to starfish.

Starfish eradication has been practiced since oyster cultivation began. Various methods are used now. Among mechanical approaches, starfish mops (figs. 16 and 17), oyster dredges (fig. 18), and suction dredges are used to remove starfish from the bottom. Recently, Bureau biologists demonstrated that several oyster enemies, including drills and starfish, may be destroyed by being buried under a comparatively thin layer of bottom mud. At present, special underwater plows are being used in experiments to develop a practical model to operate on commercial oyster beds.

Since mechanical methods of starfish control are quite expensive and only partially effective, biologists have experimented with chemicals to find those that may be practical for exterminating these predators. Good results were obtained with quicklime (calcium oxide). When quicklime is spread over the oyster beds, its falling particles imbed in the upper surfaces of the starfish body, which is covered with a delicate membrane acting as a respiratory organ. The caustic action of the lime causes surface lesions which sometimes quickly deepen and, after a few days, penetrate through the starfish body and kill it. Although many lightly injured starfish recover, the method still is quite effective if used properly.

An extremely simple method of killing starfish during transplanting of oysters from one bed to another, developed at Bureau of Commercial Fisheries Laboratory, Milford, Conn., consists of dipping dredgeloads containing oysters and other bottom material in a saturated or strong solution of common salt. This method is effective in killing not only starfish

22

Figure 15.--Group of starfish feeding on oysters. Note two empty shells in upper left corner with meats already eaten.

found together with transplanted oysters but also boring sponges, tunicates, hydroids, Crepidula, and worms. One minute of immersion, especially if the starfish are kept on deck for a short time before being returned to sea water, kills them all regardless of size.

Recently, biologists have been experimenting with certain organic chemicals mixed with dry sand or some other inert carrier and then spread on bottoms infested with starfish. Under experimental conditions some of the formulas appeared effective. This approach is being further developed and evaluated.

Several species of flatworms kill oysters, especially young ones. On the Atlantic coast two species of Stylochus, S. ellipticus and S. inimicus, commonly called oyster leech, cause heavy losses. These predators, measuring from 1/2 to 3/4 inch, crawl inside the oyster shell and feed gradually on its meat until they kill the host. According to observations made in the 1930's, these flatworms were extremely destructive in Appalachicola Bay, Fla., and other southern localities. Recently, many have been found in several areas of Long Island Sound and in isolated saltwater ponds in Martha's Vineyard, Mass.

Dipping dredgeloads of oysters in a saturated salt solution will kill these worms.

Almost all crabs with claws large and strong enough are oyster killers. In Long Island Sound, the rock crab, Cancer irroratus; green crab, Carcinides maenas; blue crab, Callinectes sapidus; several species of mud crabs belonging to the genus Neopanopeus: and other groups feed upon oysters, especially young ones with thin, brittle shells. In southern waters the stone crab, Menippe mercenaria, is abundant and destroys many oysters.

A crab usually holds an oyster in its claws and cracks off the shell piece by piece. Young oysters from a fraction of an inch to an inch or so in size are especially easy prey. Crabs are quite destructive. For example, one stone crab 3 inches wide was caged for 9 weeks with oysters from 2-2 1/2 inches long. During this period it killed 237 oysters, an average of 26 per week. The maximum number killed in a single week was 60.

Crabs can be controlled by several methods. Properly constructed wire fences surrounding oyster beds are an effective protection against crabs that cannot swim. Trapping crabs to reduce their numbers is another method.

23

Figure 16.--A typical starfish fighting boat of Long Island Sound. The boat is "mopping" starfish by dragging special mops over the oyster beds. The starfish become entangled in the mops which, after being raised, are immersed in vats of hot water to kill the pests.

Figure 17.--A closeup of a frame with 11 mops. Numerous starfish are entangled by the small rays and pincerlike structures that cover their bodies.

Figure 18.--Biologists examining a dredgeload of bottom material from an oyster bed invaded by starfish.

Recently, chemical methods of control were found efficient in some instances. For example, to protect beds of soft clams, Mya arenaria, in Maine, bait treated with chemicals poisonous to crabs was found effective. The bait usually consists of pieces of fish treated with insecticides, such as Lindane. Crabs eating this bait soon die. The chemical method, however, should be used with great caution because in certain areas, such as Chesapeake Bay, the blue crab, C. sapidus, even though an oyster predator, constitutes an important commercial fishery. In other areas, such as waters of Maine, the bait may be carried away and later eaten by lobsters, killing them. It is believed, nevertheless, that eventually poisonous baits of truly specific types will be developed so that they will affect only the undesirable species.

There is no doubt that several species of fish also feed on oysters, especially young ones. One variety, the so-called "black drum", Pogonias cromis, travels in schools and is known to destroy entire oyster beds in a comparatively short time. This fish, which attains a length of several feet and has large, strong, crushing teeth, ranges from New Jersey to Texas. As long as 40 years ago fences were used in shallow water to protect oysters from invasion and destruction by drums.

Skates also feed on oysters and clams.

Competitors

There are several species of boring sponges belonging to the genus Cliona which infest shells of oysters. Although the boring sponge is not a true parasite, because it merely excavates tunnels in the calcareous shells of oysters to provide itself a shelter, the boring sponges kill many oysters.

Probably the most common boring sponge is C. celata, which normally begins its existence in summer when it settles, after a free-swimming larval period, on shells of oysters and bores a small tunnel through the shell. Eventually the shell is honeycombed, and the sponge tunnels may penetrate the shell completely (fig. 19). Shells of old oysters severely riddled by sponges are extremely brittle. The sponge may spread over the outside of the shell and smother the oyster.

25

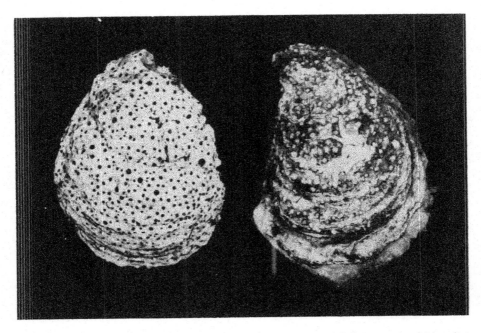

Figure 19.--Oysters affected by boring sponge (Cliona). The sponge colonies in the shell of the oyster at left have died, but the holes made by them are clearly visible. The oyster at right is still affected by vigorous, healthy colonies of sponges, portions of which are seen protruding from the shell.

The attack on oyster shells usually begins at the oldest part of the valve near the hinge. From here the sponge excavations begin to tunnel toward the shell edges. Sometimes as much as 50 percent of the shell substance may be destroyed. Eventually the sponge may attack the part of the shell where the muscle is attached and, as a result, the oyster cannot open or close its shell. At this stage the oyster usually is killed by predators, such as crabs.

Physical exhaustion of an oyster often results from its efforts to repair damages caused by a sponge. The repairs are necessary to prevent the sponge from penetrating the shell because this will expose the interior to pathogenic organisms.

Sponge damage can be reduced by maintaining oyster beds in clean condition, chiefly by removing old shells badly infested with sponges. Recently, an extremely effective treatment of sponge-infested oysters was developed by Bureau of Commercial Fisheries biologists. It is a simple procedure consisting of dipping oysters and old shells in a strong brine solution during transplanting. The brine should be a saturated solution of common salt stirred continuously. Normally, 1 minute of immersion is sufficient to kill all sponges. An additional value of the treatment is that, as already mentioned, it also destroys many other enemies and competitors of oysters.

A bivalve, a boring clam of the genus Martesia (Diplothyra), often enters the oyster shell by boring a small, round hole, and excavating a hemispherical cavity in which it spends its life, often attaining a length of three-eighths of an inch. However, this clam usually does not penetrate the shell or feed on the oyster meat and, in general, does comparatively little damage to the oyster.

Polychaete worms of the genus Polydora, usually called mud or blister worms, the majority belonging to the species P. websteri, often live on inside surfaces of oysters and cause considerable damage. When the worms are in the free-swimming stage, they enter the oysters and attach themselves to the inner surfaces of the shells close to their edges. After attachment a young worm gathers mud, placing it in a mat around itself. To protect itself from irritation caused by the mud, the oyster lays a film of new shell substance covering the mud and creating a blister. These blisters vary considerably in size from about one-half square inch to almost 1 square inch. The worm continues to live in the blister, forming a U-shaped tunnel with openings to sea water at the shell edges.

Cases are known where large numbers of Polydora infest external shell surfaces of living oysters. It has not been proven, however, that in such cases oyster mortality results directly from the Polydora attack rather than from secondary effects on weak individuals by masses of material covering and eventually smothering them.

When Polydora infect oysters in the usual way, they are able later to enlarge their burrows by edging out the calcareous shell material. The formation of blisters reduces the volume of the shell cavity and forces the oyster to spend much energy producing the secretion to cover the blisters. Heavily infested oysters are poor and watery, and mortality may be great. Nevertheless, in some instances, even heavily infested oysters possess good, healthy meats.

Polydora can be controlled successfully by dipping infected oysters in a saturated salt solution, as described for controlling sponges.

The common black mussels, Mytilus edulis, are dangerous enemies of oysters because they compete with them for space and food. Upon reaching setting size, mussel larvae, like those of oysters, descend to the bottom in search of a surface to attach to. Unlike oysters, however, they attach by a slender thread called a byssus. Sometimes the mussels set so heavily that within a short time they smother large numbers of young oysters.

Since mussels and oysters feed on the same kinds of organisms, they compete directly for food. In addition to this competitive activity, mussels deposit on oyster beds large quantities of feces and pseudofeces produced as a result of their normal feeding activities, and this often forms large masses of black, decomposing material that sometimes smother the oysters. Furthermore, if young oysters are crowded by mussels, their shells become deformed. Thus, it is virtually impossible to grow good oysters if the beds are heavily populated with mussels.

Mechanical methods for controlling mussels, such as picking them by hand or squashing them with various devices, have always been ineffective. Recently, several simple chemical methods for killing them have been developed by Bureau of Commercial Fisheries biologists and are now being further evaluated before release for general use. One method is using a material known as Victoria Blue, which kills most mussels if they are immersed in a 0.5 percent solution for 5 seconds. A better and less expensive method, however, consists of dipping mussel-infested oysters in a solution of copper sulfate. Experiments showed that this solution should be weaker than 2.5 percent. Solutions containing between 0.5 and 1.0 percent are recommended if mussels can be kept out of water for 24 hours or longer after dipping. This method is not recommended if the oysters are less than 1 inch long because considerable mortality may result.

Another serious competitor of the oyster is the slipper limpet, a snaillike animal belonging to the genus Crepidula. Most common of these is C. fornicata, sometimes found in large numbers on oyster beds of the east coast. It was introduced with shipments of American oysters to British and Dutch oyster beds which, at one time, suffered greatly from this competitor. Crepidula was also introduced to our Pacific coast with shipments of oysters from New England.

Generally, Crepidula does not interfere with fattening of oysters. However, it should be considered a dangerous oyster pest because it occupies space needed for setting and growth of oysters. As a rule, oysters and Crepidula set at about the same time. Crepidula grows much faster than the oyster and soon outgrows it. In growing, the shells of Crepidula spread over nearby oyster spat, covering them so that they suffocate and die. In one instance, examination of oyster shells collected from a bed in Long Island Sound showed seven small oysters smothered under a single Crepidula.

In Europe, Crepidula-infested oysters were dipped in a solution of corrosive sublimate. However, since corrosive sublimate is an extremely poisonous compound containing mercury, its use in contact with edible products, such as oysters, may be dangerous. Fortunately, it was found that the use of a saturated salt solution, to which reference has already been made, is extremely effective in freeing oysters of Crepidula. Affected Crepidula are usually unable to attach to a new substratum and are quickly destroyed by their enemies, such as crabs, starfish, and fish.

Anomia, often called the jingle shell, is another bivalve which competes with oysters by setting in large numbers on surfaces, such as old oyster shells, to which young oyster spat also attach. Like Crepidula, Anomia grows much more rapidly than oyster set, soon outgrowing it. In one case, 22 dead oyster spat were found under a single Anomia measuring only five-eighths of an inch in diameter.

Barnacles compete with young oysters for space and, also, to a large extent, for food and perhaps oxygen. Sometimes, when they set late in the season, barnacles cover the recently set oysters so heavily that the oysters are smothered or their shells become abnormally shaped.

No effective, practical method for barnacle control has yet been developed. It was discovered, however, that treating spat collectors with DDT prevented barnacles from setting on them. Dipping cultch in chlorinated benzenes or the mixture commonly called Polystream demonstrated that oyster spat were more numerous on treated shells and barnacles were virtually absent there in some instances, while untreated shells were heavily covered with barnacles. Treatment of cultch with Polystream also decreases setting of certain other oyster set competitors, including Crepidula.

Some varieties of bottom shrimp may be troublesome in covering the oyster beds with mud and other materials which they bring out of their burrows. As yet, such conditions have not been observed on our Atlantic coast, but on the Pacific coast two genera of ghost shrimps, Upogebia and Callianassa, cause considerable damage in this way to oyster grounds. These enemies now can be effectively controlled by spreading over infested areas dry sand or some other inactive carrier mixed with chlorinated benzenes, usually Polystream, and containing an insecticide, such as Sevin. This method, however, has not yet been approved by the U.S. Public Health Service.

Oysters, like many other sedentary bottom invertebrates, are usually found with large numbers of aquatic plants, including Ulva, Laminaria, and Zostera. Recently, a newly introduced species of algae of the genus Codium appeared on some beds in New England. Often a thick growth of these plants interferes with circulation of the water, thus depriving oysters of their food. When the heavy growth of algae dies, it smothers the oysters and, in decomposing, deprives them of necessary oxygen.

OYSTER INDUSTRY

The United States leads all countries in the quantity of oysters grown for market and in value of the product. The Eastern oyster represents about 80 percent of the total production. At present, the Chesapeake Bay area leads the sections producing Eastern oysters, while the Gulf of Mexico occupies second place (table 1). Several thousand men are employed in various aspects of shellfisheries, ranging from people who cultivate oysters to those who open them and pack their meats for market.

Oyster farming has been practiced in the West since the days of the Romans, and oyster cultivation was practiced in China long before the Christian Era. In North America, Indians living near the ocean knew the value of oysters and other mollusks and used them widely as food. Nevertheless, farming, or, as it is often called, cultivation, of oysters on the eastern

Table 1.--Production and value of Eastern oysters - by States and areas, 1956-60

Area	1960		1959		1958		1957		1956	
	Thousand pounds	Thousand dollars	Thousand pounds	Thousand dollars	Thousand pounds	Thousand dollars	Thousand pounds	Thousand dollars	Thousand pounds	Thousand dollars
Maine	3	1	4	2	4	2	6	4	---	---
Massachusetts	111	153	118	163	113	137	152	152	229	200
Rhode Island	25	23	6	4	3	3	3	4	31	25
Connecticut	361	448	259	285	156	177	244	216	246	211
New York	810	905	890	908	1,057	987	1,067	1,019	1,070	1,001
Total-New England	1,310	1,530	1,277	1,362	1,333	1,306	1,472	1,395	1,576	1,437
New Jersey	167	161	207	190	829	675	2,720	1,782	5,503	3,023
Delaware	177	119	295	158	2,410	1,717	4,194	2,227	1,893	783
Total-Delaware Bay area	344	280	502	348	3,239	2,392	6,914	4,009	7,396	3,806
Maryland	11,771	8,426	11,966	7,233	12,026	6,668	14,732	7,601	15,843	8,792
Virginia	15,340	10,884	21,356	13,374	25,504	14,127	20,090	9,848	21,221	9,900
Total-Chesapeake Bay area	27,111	19,310	33,322	20,607	37,530	20,795	34,822	17,449	37,064	18,692
North Carolina	1,216	560	1,311	587	1,041	434	1,086	479	1,318	564
South Carolina	2,627	920	1,918	379	1,437	288	1,845	370	2,186	442
Georgia	232	59	248	61	143	35	112	28	120	36
E. Coast Florida	44	13	39	12	30	8	26	7	32	7
Total-South Atlantic	4,119	1,552	3,516	1,039	2,651	765	3,069	884	3,656	1,049
W. Coast Florida	1,931	483	1,415	405	795	218	710	199	857	206
Alabama	1,169	317	895	278	458	111	1,291	288	769	174
Mississippi	2,391	535	333	82	579	123	863	186	846	173
Louisiana	8,311	2,304	9,667	2,646	8,265	2,426	10,490	2,756	10,056	2,238
Texas	2,296	655	1,411	396	311	119	953	262	985	286
Total-Gulf	16,098	4,294	13,721	3,807	10,408	2,997	14,307	3,691	13,513	3,077
Grand total	48,982	26,966	52,338	27,163	55,161	28,255	60,584	27,428	63,205	28,061

coast began only about one hundred years ago.

Oyster farming usually consists of several operations, depending upon local conditions. In the Southern States, oysters are gathered principally from State-owned, or public, bottoms, where little or no cultivation is done. In the North, however, especially in Long Island Sound and the waters of New York State, oysters come almost entirely from privately owned and leased grounds, where extensive cultivating operations are a general practice.

Normally, before planting shells to collect the new generation of oysters, oyster farmers clean the beds to remove as much debris as possible and simultaneously destroy such enemies as starfish and drills. After the bottom is clean, and just before the larvae set, oyster and clam shells are placed on the bottom to "catch" the new generation of oysters. The shells and other material used for this purpose are called "cultch".

The set, collected on beds that may be damaged later in the season by winter storms or invaded by enemies, is dredged a few months after setting and transplanted to deeper, safer waters where young oysters are left to grow until they reach marketable size. If oysters are too numerous and grow too rapidly, they may be transplanted several times to new grounds to offer them sufficient space for growth. Normally, after cultivated oysters reach 3 or 4 inches they are transplanted to fattening grounds where, because of favorable conditions, they can store a large amount of glycogen in their meats and acquire the desired flavor and appearance. Finally, they are dredged again and, after culling, grading, and packing, are shipped to market.

Production of seed oysters constitutes an important part of the oyster industry. These activities are probably best developed in New England, especially along the northern shore of Long Island Sound, where many companies produce oyster seed. By New England standards, seed oysters are those which set in summer and are offered for sale either during the following fall or the next spring. In other areas, however, for example, in Chesapeake Bay, the term "seed oyster" may apply to all oysters smaller than 3 inches taken from public grounds. Seed oysters are sold to oyster cultivators who plant them on privately owned or leased bottoms.

Oysters are fished by a variety of methods, including such primitive ways as handpicking, and progressing to use of tongs (fig. 20), mechanical dredges, suction dredges, and escalator dredges.

Handpicking is practiced on public grounds in Southern States where oyster flats are often exposed to low water stages. Grabs, or hooks, are sometimes used. The use of manual tongs is limited to comparatively shallow water but is profitable, nevertheless, on grounds where oysters are plentiful. In the James River, Va., where no mechanical methods of fishing are

Figure 20.--An oysterman tonging oysters on public grounds in Chesapeake Bay. The tongs are operated by scissorlike motion.

permitted on natural beds, a tonger may catch sometimes 25 to 30 bushels a day.

Dredges vary in methods of their operation and in their holding capacity. In some areas, such as the public grounds of Chesapeake Bay and Connecticut, no powered craft can be used and only hand-operated dredges are permitted (fig. 21). Therefore, these dredges are comparatively small, having a capacity of only about 1 or 2 bushels. On private grounds, however, large machine-hoisted dredges are used (figs. 22 and 23). The capacity of these dredges may exceed 20 bushels. Normally, a typical oyster dredging boat of Long Island Sound is equipped with two dredges, one on each side (fig. 24). In the past, some of the largest vessels, equipped with three dredges on each side, were capable of harvesting at the rate of 1,400 bushels per hour.

After World War II, considerable progress was made in mechanizing and improving methods of oyster harvesting. One step in this direction was constructing a suction dredge, which works on the principle of a vacuum cleaner (fig. 25). Suction is produced by a powerful jet of water, and the power of the suction carries oysters and other mate-rials from the bottom to the conveyor located on deck of the dredge boat. While moving on the conveyor, the material is examined and handled as necessary, including sorting of oysters according to size.

In addition to harvesting oysters, suction dredges are extremely efficient in clearing oyster bottoms of various enemies, such as starfish, mussels, crabs, Crepidula, and sometimes drills. A larger suction dredge vessel of this type has been in almost continuous operation in Long Island Sound since 1948. It is about 100 feet long, 30 feet wide, and draws 8 feet of water. Its capacity is 10,000 bushels. On the basis of its past performance, it is regarded as an extremely efficient and versatile device.

The escalator or, as it is sometimes called, scooper-type of harvester also has been developed since the war and is used in several oyster-producing areas. This type of harvester is most effective in relatively shallow water. The dredge of a scooper-type harvester has a rakelike appearance and is equipped with steel teeth. During operation it is lowered to the bottom where it rests on special runners on each side which prevent the dredge from

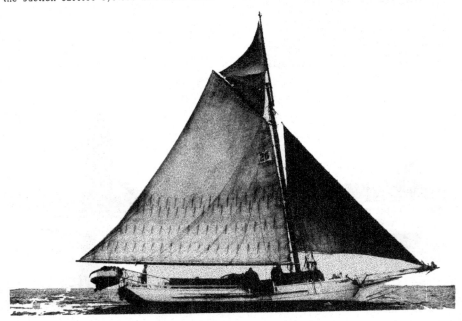

Figure 21.--Sailboat dredging oysters on public beds.

30

Figure 22.--Conventional type of machine-operated oyster dredge widely used on private oyster grounds.

Figure 23.--Modern type of oyster dredge which is easily unloaded by opening its bottom.

Figure 24.--Closeup of two oyster dredges designed to operate on each side of the boat.

digging too deep. The material culled by the rake is scooped by flexible hoops and brought to the deck on an endless chain-type conveyor.

The escalator hydraulic harvester is still another type of oyster dredging gear now being developed by Canadians. It is reported to be quite effective in comparatively shallow water because it picks up most oysters encountered on its way. Presumably it causes less bottom disturbance, or damage, than other types of dredges of the same size. It operates by using strong jets of water to lift oysters from the bottom and to place them on an endless belt which brings them to the surface. The experimental models of this type of dredge can bring up about 50 bushels per hour.

It is estimated that in territorial waters of the United States there are about 1,400,000 acres of bottom designated as oyster-producing areas. Of this entire acreage only about 185,000 acres are privately leased or controlled. Even though a considerable area of these privately controlled grounds is not cultivated, these beds produce, nevertheless, about 50 percent of the total oyster crop of the United States. Therefore, there is a great difference in productivity between privately controlled and public oyster beds.

Federal authorities have no jurisdiction over shellfisheries in State waters. Leasing or purchasing of oyster bottoms must be accomplished through fisheries authorities of the State in which they are located. In many States, however, most of the best oyster bottoms are regarded as public oyster beds and, therefore, cannot be cultivated by private interests. Instead, these areas are theoretically open for fishing to all citizens of the State, who, however, do not directly participate in their cultivation. As a result, because of bad management, many formerly prolific public oyster beds have been depleted to such an extent that in several States the oyster industry virtually no longer exists, while in others, such as the Chesapeake Bay States, productivity has generally declined.

Depletion of public oyster grounds can be illustrated by the fate of shellfisheries in several States where private oyster farming is not encouraged. Perhaps the best example is the State of Georgia where, according to the U.S. Coast and Geodetic Survey, there were, at the turn of the century, about 30,000 acres suitable for oyster cultivation but which were set aside as public grounds. In 1908, this area produced about 1,446,100 bushels of oysters; in 1923, production had dropped to 245,762 bushels, and in 1939, to 78,133 bushels. The decrease in oyster production could not be ascribed to outside causes, such as floods or dredging. The chief cause of this decline

Figure 25.--Large suction dredge operating on oyster beds in Long Island Sound and in the waters of the State of New York. Description in text.

probably was a lack of constructive management. This deficiency led gradually to destruction of the unusually productive oyster beds.

Depletion of natural oyster beds, or public grounds, can be demonstrated further by another example. In some sections of the Potomac River where, not long ago, oysters were abundant, the grounds are now so exhausted because of overfishing and lack of cultivation that production of oysters averages less than 1 bushel per acre. Yet, by properly cultivating and protecting oysters, an acre of oyster bed can support from 500 to 1,200 bushels. If State shellfisheries authorities would follow the practices of private oyster cultivators, they could restore the oyster populations on the depleted grounds.

SANITARY CONTROL

Oysters can accumulate certain pathogenic bacteria harmful to man. Because of this, the U.S. oyster industry, as other shellfisheries, is subject to strict sanitary regulations. Sanitary supervision is exercised by the health departments of different States, by the U.S. Public Health Service, and by the Food and Drug Administration.

Waters from which oysters are taken to be sold for food are inspected and tested by health authorities. These organizations also guard against unlawful adulterations of oyster meats. After oyster meats are removed from the shells, their purity is determined by rigid bacteriological examination and, in instances of chemical pollution, by other tests. Oysters failing to meet required standards cannot be released for sale.

Shucking houses (where oysters are opened) and all equipment coming in contact with oyster meats must be kept clean. Furthermore, all employees who open oysters must undergo periodic medical examinations to ascertain that they are not suffering from contagious diseases which they may transfer to others by handling oysters.

Oysters from contaminated areas may be purified in special tanks through which large quantities of pure water are passed. The water

33

for this purpose is prepared by several methods, including chlorination and use of ultraviolet rays, which destroy bacteria. In a short time, oysters held in clean water expel ingested bacteria and then are safe to eat. Such methods of purification, or depuration, of oysters from contamination waters are used quite successfully in Holland, England, and France, but are not yet generally accepted in the United States, although there is a trend in this direction.

As the population of the United States expands, more coastal waters will be used for disposal of waste, reducing the clean areas now available for shellfish culture. It seems logical, therefore, to adopt procedures which will provide the public with oysters that are safe to eat. Perhaps in the near future depurated oysters will be as well accepted as pasteurized milk.

SELECTED REFERENCES

General

BAUGHMAN, J. L.
  1948. An annotated bibliography of oysters, with pertinent material on mussels and other shellfish and an appendix on pollution. Texas A. & M. Research Foundation, College Station, Tex., 794 p.
CHURCHILL, E. P., Jr.
  1920. The oyster and the oyster industry of the Atlantic and Gulf coasts. [U.S.] Bureau of Fisheries, Report of the Commissioner of Fisheries for the fiscal year 1919, Appendix 8 (Document 890), 51 p.
GALTSOFF, PAUL S.
  1965. The American oyster Crassostrea virginica Gmelin. U.S. Fish and Wildlife Service, Fishery Bulletin, vol. 64, iv + 480 p.
KORRINGA, P.
  1952. Recent advances in oyster biology. Quarterly Review of Biology, vol. 27, no. 3, p. 266-308, 339-365.
YONGE, C. M.
  1960. Oysters. Willmer Brothers & Haram Ltd., London, 209 p.

Anatomy and Physiology

DAVIS, H. C.
  1958. Survival and growth of clam and oyster larvae at different salinities. Biological Bulletin, vol. 114, no. 3, p. 296-307.
DAVIS, HARRY C., and PAUL E. CHANLEY.
  1956. Spawning and egg production of oysters and clams. Biological Bulletin, vol. 110, no. 2, p. 117-128.

DAVIS, HARRY C., and ROBERT R. GUILLARD.
  1958. Relative value of ten genera of micro-organisms as foods for oyster and clam larvae. U.S. Department of the Interior, Fish and Wildlife Service, Fishery Bulletin 136, vol. 58, p. 293-304.
LOOSANOFF, V. L.
  1952. Behavior of oysters in water of low salinites. National Shellfisheries Association 1952 Convention Addresses, p. 135-151.
  1958. Some aspects of behavior of oysters at different temperatures. Biological Bulletin, vol. 114, no. 1, p. 57-70.
  1962. Effects of turbidity on some larval and adult bivalves. Gulf and Caribbean Fisheries Institute, Proceedings of the 14th Annual Session, November 1961, p. 80-95.
LOOSANOFF, VICTOR L., and HARRY C. DAVIS.
  1952. Temperature requirements for maturation of gonads of northern oysters. Biological Bulletin, vol. 103, no. 1, p. 80-96.
LOOSANOFF, VICTOR L., and JAMES B. ENGLE.
  1940. Spawning and setting of oysters in Long Island Sound in 1937, and discussion of the method for predicting the intensity and time of oyster setting. [U.S.] Bureau of Fisheries, Bulletin No. 33, vol. 49, p. 217-255.
  1947. Effect of different concentrations of micro-organisms on the feeding of oysters (O. virginica). [U.S.] Fish and Wildlife Service, Fishery Bulletin 42, vol. 51, p. 31-57.
NELSON, T. C.
  1938. The feeding mechanism of the oyster. I. On the pallium and the branchial chambers of Ostrea virginica, O. edulis, and O. angulata. Journal of Morphology, vol. 63, no. 1, p. 1-61.
YONGE, C. M.
  1926. Structure and physiology of the organs of feeding and digestion in Ostrea edulis. Journal of Marine Biological Association, vol. 14, no. 2, p. 295-386.

Artificial Cultivation

LOOSANOFF, V. L.
  1956. On utilization of salt water ponds for shellfish culture. Ecology, vol. 37, no. 3, p. 614-616.
LOOSANOFF, V. L., and HARRY C. DAVIS.
  1963a. Rearing of bivalve mollusks. In F.S. Russell (editor), Advances in marine biology, vol. 1, p. 1-136. Academic Press, London and New York.

LOOSANOFF, V. L., and HARRY C. DAVIS.
1963b. Shellfish hatcheries and their future. [U.S.] Fish and Wildlife Service, Commercial Fisheries Review, vol. 25, no. 1, p. 1-11.

Enemies and Their Control

BUTLER, PHILIP A.
1953. The southern oyster drill. National Shellfisheries Association 1953 Convention Papers, p. 67-75.
CARRIKER, MELBOURNE ROMAINE.
1955. Critical review of biology and control of oyster drills, Urosalpinx and Eupleura. U.S. Department of the Interior, Fish and Wildlife Service, Special Scientific Report--Fisheries No. 148, 150 p.
GALTSOFF, PAUL S., and VICTOR L. LOOSANOFF.
1939. Natural history and method of controlling the starfish (Asterias forbesi, Desor). [U.S.] Bureau of Fisheries, Bulletin No. 31, vol. 49, p. 75-132.
LAIRD, MARSHALL.
1961. Microecological factors in oyster epizootics. Canadian Journal of Zoology, vol. 39, no. 4, p. 449-485.
LANDAU, HELEN, and PAUL S. GALTSOFF.
1951. Distribution of Nematopsis infection on the oyster grounds of the Chesapeake Bay and in other waters of Atlantic and Gulf States. Texas Journal of Science, vol. 3, no. 1, p. 115-130.
LOOSANOFF, VICTOR L.
1958. New method for control of enemies with common salt. [U.S.] Fish and Wildlife Service, Commercial Fisheries Review, vol. 20, no. 1, p. 45-47.
1961. Recent advances in the control of shellfish predators and competitors. Gulf and Caribbean Fisheries Institute, Proceedings of the Thirteenth Annual Session, November 1960, p. 113-127.
LOOSANOFF, VICTOR L., and JAMES B. ENGLE.
1942. Use of lime in controlling starfish. [U.S.] Fish and Wildlife Service, Research Report 2, 29 p.
MacKENZIE, CLYDE L., Jr.
1961. A practical chemical method for killing mussels and other oyster competitors. U.S. Fish and Wildlife Service, Commercial Fisheries Review, vol. 23, no. 3, p. 15-19.
MACKIN, J. G.
1960. Status of researches on oyster disease in North America. Gulf and Caribbean Fisheries Institute, Proceedings of the Thirteenth Annual Session, November 1960, p. 98-109.

MACKIN, J. G., P. KORRINGA, and S. W. HOPKINS.
1952. Hexamitiasis of Ostrea edulis L. and Crassostrea virginica (Gmelin). Bulletin of Marine Science of the Gulf and Caribbean, vol. 1, no. 4, p. 266-277.
MENZEL, R. WINSTON, and FRED E. NICHY.
1958. Studies of the distribution and feeding habits of some oyster predators in Alligator Harbor, Florida. Bulletin of Marine Science of the Gulf and Caribbean, vol. 8, no. 2, p. 125-145.
RAY, S. M., and A. C. CHANDLER.
1955. Dermocystidium marinum, a parasite of oysters. Experimental Parasitology, vol. 4, no. 2, p. 172-200.
STAUBER, LESLIE A.
1945. Pinnotheres ostreum, parasitic on the American Oyster, Ostrea (Gryphaea) virginica. Biological Bulletin, vol. 88, no. 3, p. 269-291.

Oyster Industry

ANONYMOUS.
1948. Mechanization of oyster cultivation. Part 1--Recent developments and improvements in oyster dredges. [U.S.] Fish and Wildlife Service, Commercial Fisheries Review, vol. 10, no. 9, p. 12-16.
LEE, CHARLES F., and F. BRUCE SANFORD.
1963. Oyster industry of Chesapeake Bay, South Atlantic, and Gulf of Mexico. U.S. Fish and Wildlife Service, Commercial Fisheries Review, vol. 25, no. 3, p. 8-17.
MEDCOF, J. C.
1961. Oyster farming in the Maritimes. Fisheries Research Board of Canada, Bulletin No. 131, 158 p.
U.S. DEPARTMENT OF HEALTH, EDUCATION, and WELFARE, PUBLIC HEALTH SERVICE.
1957. Sanitary control of the shellfish industry. Part II: Harvesting and processing, p. 1-26.

Olympia Oyster, Ostrea lurida

GALTSOFF, PAUL S.
1929. Oyster industry of the Pacific coast of the United States. [U.S.] Bureau of Fisheries, Report of the Commissioner of Fisheries for the fiscal year 1929, appendix 13 (Document 1066), p. 367-400.
HOPKINS, A. E.
1937. Experimental observations on spawning, larval development, and setting in the Olympia oyster, Ostrea lurida. [U.S.] Bureau of Fisheries, Bulletin No. 23, vol. 48, p. 439-503.

Pacific or Japanese Oyster,
Crassostrea gigas

CAHN, A. R.
    1950. Oyster culture in Japan. U.S. Fish and Wildlife Service, Fishery Leaftlet 383, 80 p.
GLUD, JOHN B.
    1949. Japanese methods of oyster culture. U.S. Fish and Wildlife Service, Commercial Fisheries Review, vol. 11, no. 8, p. 1-7.
KINCAID, TREVOR.
    1951. The oyster industry of Willapa Bay, Washington. Calliostoma Company, Seattle, Wash., 45 p.
QUAYLE, D. B.
    1956. Pacific oyster culture in British Columbia. Provincial Department of Fisheries, Victoria, B.C., 33 p.
STEELE, E. N.
    1964. The immigrant oyster (Ostrea gigas) now known as the Pacific oyster. Warrens Quick Print, Olympia, Wash., 179 p.

European oyster, Ostrea edulis

KORRINGA, P.
    1940. Experiments and observations on swarming, pelagic life, and setting in the European flat oyster, Ostrea edulis L. Archives Neerlandaises de Zoologie, vol. 5, p. 1-249.
LOOSANOFF, V. L.
    1955. The European oyster in American waters. Science, vol. 121, no. 3135, p. 119-121.
ORTON, J. H.
    1937. Oyster biology and oyster-culture. Arnold, London, 211 p.

MS. #1379

CPSIA information can be obtained
at www.ICGtesting.com
Printed in the USA
LVOW13s1254180517
535003LV00017B/421/P